Observations on the Quantum Mechanical Nature of Gravity

Greg Feild

October 8, 2016

About the author: by Greg Feild

I earned a Ph.D in experimental high energy physics from the Pennsylvania State University working on HERA at DESY in Hamburg, Germany studying photoproduction and deep inelastic scattering in electron-proton collisions.

I did my postdoctoral studies with Yale University working at Fermilab on the CDF experiment at the Tevatron. My primary research interest was particle hadronization in charmonium production in proton-antiproton collisions.

I am retired now but still thinking about things as my two dogs allow.

Dedication:

This book is dedicated to the experimental physicists, without whom there would be nothing to theorize about, and to my nephews; Knox, Zac, and Max, without whom there would not be this theory nor this book.

Abstract:

This paper explores and develops the ideas presented in "A Quantum Mechanical Theory of Gravitational Interactions".

A satisfactory unification of the four fundamental forces can be achieved if we assume the electron neutrino is a point magnetic charge corresponding to the point electric charge of the electron. This is not to be confused with the idea of the classical magnetic monopole. The neutrino is, however, assumed to have a magnetic moment. This leads to the conclusion that magnetic fields are caused by the motion of mass rather than electric charge.

The predicted value of the magnetic moment of the electron neutrino is

$$\mu_v = (1.61 \times 10^6) \cdot \mu_B$$

where μ_B is the classical Bohr magneton.

Introduction:

Gentle readers,

My previous 'book', "A quantum mechanical theory of gravitational interactions", was originally intended for publication in Physical Review Letters; hence its brevity and style. I felt it fit their criteria rather nicely, but I was mistaken. It took a creative use of white space and some filler to turn this journal letter into a 24 page book.

However, the extra time gave me an opportunity to think more on some of the ideas and add extra content in the form of an addendum. Still, the 'first edition' of my book contained several silly and embarrassing errors. I was able to correct these mistakes fairly seamlessly in this brave new world of online self-publication.

The drawbacks, of course, were that I was; author, editor, proofreader, fact checker, math checker, reviewer and my own biggest fan! So, of course, mistakes were made.

Given the brevity of the previous book, some of the assumptions and many of the implications were not explicitly spelled out or fully developed. Indeed, I will say that I was not aware of some of the implications at the time nor inclined to work them out as math is not my strongest suit.

As my previous book, this volume is meant for fellow (working!) physicists who are certainly more educated than I on many details, including the manipulation of the units of high energy physics theory and general relativity, and are better equipped, and essential, for working through all the mathematics. I hope people will find at least some of these book's conclusions palatable and will be willing to invest their time and energy in helping work these ideas through.

The past two months of googling myself, theoretical papers that may support my thesis, and/or newly released or reissued books on the subjects of gravity and QCD, have shown me that several of the ideas in my first book have been proffered before in various integration or theory reformulation schemes. I say such overlapping theories are to be expected, exciting and encouraging to see. I also say;

Keep it simple!
Physics is fun!

Greg F.

In "A quantum mechanical theory of gravitational interactions" (QMOGI), we derived/deduced a quantum theory of gravity where the mass of a particle was equivalent to its gravitational charge. In an attempt to assign a gravitational charge to the photon (as well as the graviton itself) we concluded that the gravitational charge of a particle was not a constant of the particle's motion, but dependent on its energy. Particles with a gravitational charge were determined to interact by the exchange of a graviton of mass=0, spin=1.

In developing this theory, we did not assume that it had to be compatible with, or reduce to the general theory of relativity. In fact, by considering a simple hypothetical example of a two neutrino bound state, we found our quantum mechanical theory of gravity to disagree with general relativity.

Let us now make explicit the conclusion that the general theory of relativity is incorrect. As we argued in QMOGI, this is because general relativity treats mass and energy as 'separate, but equal', when in fact they are equivalent.

Our new quantum mechanical theory subsumes any additional energy gained by a particle during an interaction into the particle's mass (a.k.a. gravitational coupling charge).

A simple conclusion of this quantum theory, and one that should be easy to check, is that the deflection of starlight by massive bodies would be due to interactions with the gravitational charge of the constituent photons.

If we have cast doubt on the general theory of relativity; then concerns such as acceleration, inflation, dark matter, dark energy, etc., become moot.

In addition, if we have recovered three dimensional space and Newtonian time; and retained the big bang, then we now have a place for the fabled inertial coordinate system of yesteryear; R=0 at the origin of the 'original singularity' that started it all! (We also should now be able to determine our position, R_earth, by gazing 'inward' and 'outward' and comparing the relative mass densities, gravitational redshifts, and/or measurements of the relative brightness of familiar celestial objects.

The rest of this paper will be devoted to some other matters raised in QMOGI, or to those that now suddenly seem explicable. Broadly, these will be;

1) the $1/R^2$ law, potential energy, and 'action at a distance'!
2) missing mass
3) gravitational redshift
4) grand unification

Gravity:

In QMOGI, we determined that a graviton of spin=1 is the exchange particle mediating the gravitational interaction. This graviton is massless yet carries a gravitational charge proportional to its energy. As we know, the graviton must be massless as the force of gravity follows the $1/R^2$ law and is, in principle, of infinite reach.

Now that we have determined a massless graviton as the mediator of the gravitational force, we can deduce the $1/R^2$ law from the uncertainty principle.

The virtual nature of the graviton as an exchange particle means it must satisfy the uncertainty relationship;

$$\Delta E \cdot \Delta t = \hbar \qquad (1)$$

We can calculate Delta.t from

$$\Delta t = R/c \qquad (2)$$

where R is the distance between any two interacting particles and c is the speed of light. We can now solve for Delta.E;

$$\Delta E = \hbar c / R \qquad (3)$$

If we consider Delta.E to be the gravitational potential energy between the two particles, then the classical force between the two particles will go as $1/R^2$.

Ironically, perhaps, in QMOGI we concluded that the potential energy, while a useful concept, is not really a fundamental type of energy and should be removed from the pantheon.

Concerning action at a distance; it is well known that Newton and his contemporaries found it quite unacceptable, yet they seem to have not been as equally worried about the instantaneous transmission of this same force. One can understand why. As the earth has been revolving around the sun for hundreds of millions of years, it seems trivial to worry that it may have "originally" taken eighteen minutes for them to couple up. I speak of them as coupled as it is also hard to imagine each and every particle on the earth and the sun constantly generating and exchanging "new" gravitons. How would they even know where to go as their target is moving and millions of miles away?

I believe a more natural picture of this model is that all particles are constantly coupled via the graviton/gravitational field and are exchanging energy and momentum in accordance with Newton's third law. Additionally, for us to be interacting with the Andromeda galaxy, for example, I believe we have to assume that the original gravitational coupling was initiated long ago when all the particles were much closer together. In other words, I don't think particles send out feeler gravitons 'on spec' looking for something to couple.

But, back to physics.

Missing mass:

In QMOGI, we replaced QCD with gravity, and quarks with leptons. In fact, we postulated there are only two fundamental particles; the electron and the electron neutrino.

In the beginning, electrons and neutrinos were generated in equal numbers along with their corresponding antiparticles. These four fundamental particles combined to produce the four particles familiar to us today; the proton, the neutron, the electron and the electron neutrino. The protons, neutrons, and electrons combined to form the atoms leaving the neutrinos alone; footloose and fancy free.

We conclude the universe is filled with electron neutrinos the number of which being equivalent to the estimated number of protons in the world.

Gravitational redshift:

Since we no longer consider space to be a physical substance capable of expanding and contracting (it is now gone with the ether), we must assume that the gravitational redshift of light from a distant galaxy is due to the gravitational drag of the mass of the galaxy on the emitted photons. The photons are now considered to have a gravitational charge proportional to their energy since we believe mass and energy are equivalent.

The universe is no longer flying apart but we can still use the gravitational redshift of light to measure the distance of a faraway galaxy by calculating the gravitational work done on a photon by the mass of the emitting galaxy.

Grand unification:

In the current standard model, electromagnetism and the weak interaction are unified by electroweak theory. These two forces are unified in the sense that the same mechanism is responsible for the generation of the photon and the massive weak exchange bosons; the W and Z. In addition, the forces are of equivalent "strength" (equally likely to facilitate interactions) when the masses of the particles involved are on the order of the mass of the Z.

Still, this unification is unsatisfactory in the sense that we still have two separate coupling constants; the electric charge and the weak charge.

In QMOGI (my apologies!), we determined the weak charge of a particle was equivalent to its mass. In addition, we concluded the color charge of QCD was also a particle's mass and QCD was just a manifestation of gravity.

So, in this revised standard model the four forces have been united into electromagnetism and gravity, and there are now only two fundamental particles; the electron and the electron neutrino.

There is still a problem of symmetry. The electron has mass and electric charge while the neutrino has only mass. There are 3 possible solutions to this conundrum;

1) the neutrino has a tiny electric charge
2) electric charge is some kind of 'emergent' property of mass
3) the neutrino is a magnetic point charge

The most beautiful answer is usually the winner, so we will chose number three.

If the neutrino is a magnetic point charge complementary to the electric point charge of the electron, then it should certainly have a magnetic moment. Since we are trying to unify physics we will assume the magnetic moment is derived in the usual way;

$$\mu_v = e \hbar / (2 m_v c) \quad (4)$$

$$\mu_v = (m_e/m_v) \mu_B \quad (5)$$

where e is the electric charge, m_v is the mass of the neutrino, m_e is the mass of the electron and μ_B is the classical Bohr magneton.

As the neutrino has no electric charge we must assume that the neutrino's magnetic moment is due to its 'spinning mass' and that the factor of e is an historical artefact.

This would mean that magnetic fields are caused by the motion of mass and not electric charge.

Conclusion:

This theory could be tested by investigating whether a neutrino beam generates a magnetic field.

References:

"Elementary Modern Physics", Second Edition
Richard T. Weidner and Robert L. Sells
College Physics Series/Allyn and Bacon, 1968

"Introduction to High Energy Physics", Third Edition
Donald H. Perkins
Addison-Wesley Publishing Company, 1987

"Gravitation and Cosmology"
Steven Weinberg
John Wiley and Sons, 1972

Wikipedia

Book shelf:

A Survey of Physical Theory   Max Planck

Physics and Philosophy   Werner Heisenberg
The Physical Principles of the Quantum Theory   Werner Heisenberg

The Structure of Scientific Revolutions   Thomas Kuhn
How We Got to Now   Steven Johnson
To Explain the World   Steven Weinberg

The Origins of Knowledge and Imagination   Jacob Bronowski
Science and Human Values   Jacob Bronowski

Faraday, Maxwell and the Electromagnetic Field   Nancy Forbes and
                                                        Basil Mahon

everything:

    quantum loops
    wisps o' mind
    beauty clouds peace
    everything, mystery
    words, feelings say
    meaning, cyclic,
    eternal recurrences,
    triangle, Platonic shadows
    all is circle

the time worn threshold:

    the time worn threshold
    flecked and freckled
    grainy, grey
    tales of feet well-traveled
    ruts of remembrance;
    tacit first steps, and
    stubbing toes
    acquaintances unfamiliar
    foreign, foe, and formal
    friends
    a welcoming runnel
    running right up to the split
    in the sun-warped linoleum

listening to prozac:

    it be.
    you willed that,
    did you not ?
    retool
    the ghost in the machine
    process, procedure,
    pedagogy, purgatory
    primary
    purged of self
    somatic switches
    sidled to the void
    null space
    synaptic fluids
    sloshing with preordination
    wet, wonderful world

rapture:

    all will be,
    in the end,
    raised up;
    heavenward
    as is their wont.
    air buoys
    water drowns
    out cries for the world
    to cast of its caul.
    Ezekiel's wheel;
    torrid, turning
    toroidal coils
    of cold fusion.
    frisson, frayed wires
    magnetic rings around
    the collar of earth's pole;
    pushing, pulling
    inexorably
    toward equilibrium
    revolutionary stasis and
    revelation
    objects linger lout-like
    on the black periphery
    of the horizon and the abyss
    refusing to give up the ghost

Notes:

www.ingramcontent.com/pod-product-compliance
Lightning Source LLC
Chambersburg PA
CBHW070342190526
45169CB00005B/2015